BAHAMIAN BOAS

A TABBY TALE

BY ELAINE A POWERS

ILLUSTRATED BY ANDERSON ATLAS

LYRIC POWER

ISBN: 978-0-9991669-5-6
BISAC Category * Juvenile NonFiction / Nature / Snakes
For sales and Distribution Contact Elaine A. Powers at iginspired@gmail.com

My name is Tabby, I'm a Five-Finger Fairy
I have friends you might think are scary.
My tale is about some snakes you should know,
The amazing, native Bahamian boa.

4

Snakes often have bad reputations
In the Bahamas and many nations.
People think snakes will cause them great harm
Merely because they don't have any arms.
Some people fear them because of religion;
So learning about snakes is the place to begin.

5

All snakes are reptiles without any legs,
They're carnivores, have scales, breathe air, and make eggs.

The five kinds of Bahamian snakes
are quite calm.
The largest, boas or fowl snakes,
live in gammalemme and palm.

Caribbean boas evolved into many;
Some islands have boas, but others don't have any.
The islands are all separated by sea,
So it's not a surprise that different kinds came to be.

8

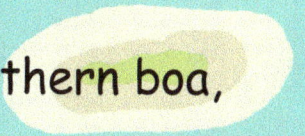

There's the Northern, the Central and the Southern boa,
A new one, the Conception Bank Silver boa.
The lost Crooked-Acklins boa was recently found,
Boa discoveries continue to abound!
They're called endemic, because they only live here;
Which means, of course, that they're really quite rare.

Boas will usually live on their own,
Dwelling in forests of their Bahamian home.
Boas are arboreal, living in the trees,
But you might find them underneath rocks and leaves.
Some boa species are rare, hard to find.
To very specific forests, they are confined.

Though boas need forests to live and find prey,
They need places to hide when danger comes their way.
Tree holes, rotten logs, and rocks are good places,
Providing the boas with their safety spaces.

Some call the Bahamian boa "rainbow,"
Due to the scales' sunlit iridescent glow.
Each species has a unique pattern over
    grey-brown.
Their shining beauty is of great renown.
The special structures, called
    iridophores,
Create quite of splash of colors galore.
On scale tops, like gloss, are layers of
    crystal;
For moving, this very smooth skin is
    beneficial.

The microridges reduce friction for gaining speed
And repel water - in the tropics an important need.
In being called Bahamian boa there is no shame
But perhaps "Rainboa" would be a better name.

13

Boas, like all reptiles, are called ectotherm,
A fancy term for "cold-blooded" that all should learn.
Snakes depend on the environment to keep
themselves warm,
Using sun and shadow, their temps to transform.

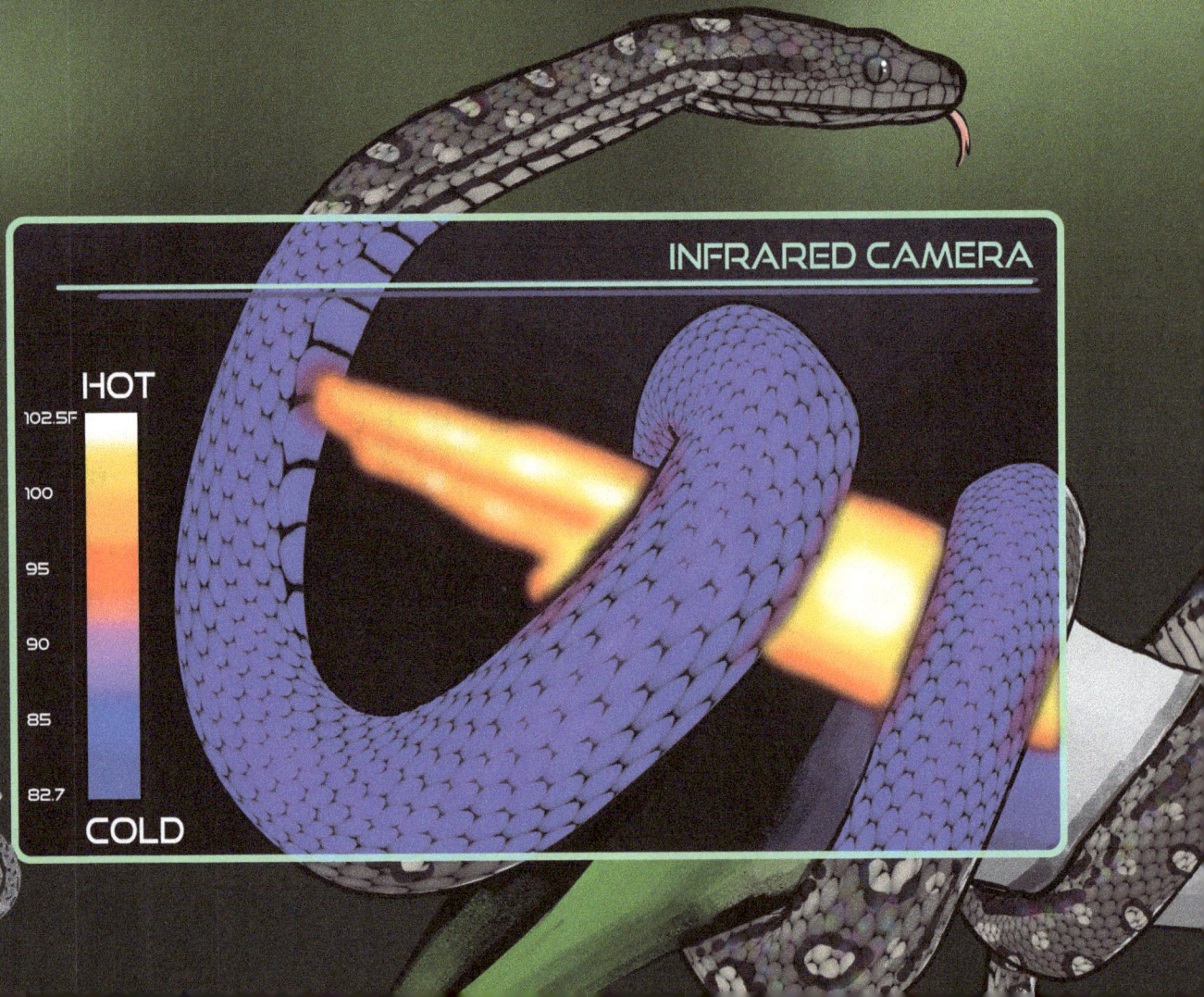

INFRARED CAMERA

HOT

102.5F

100

95

90

85

82.7

COLD

14

Boas don't have ears and can't really "hear,"
But they sense sound vibrations when someone is near.

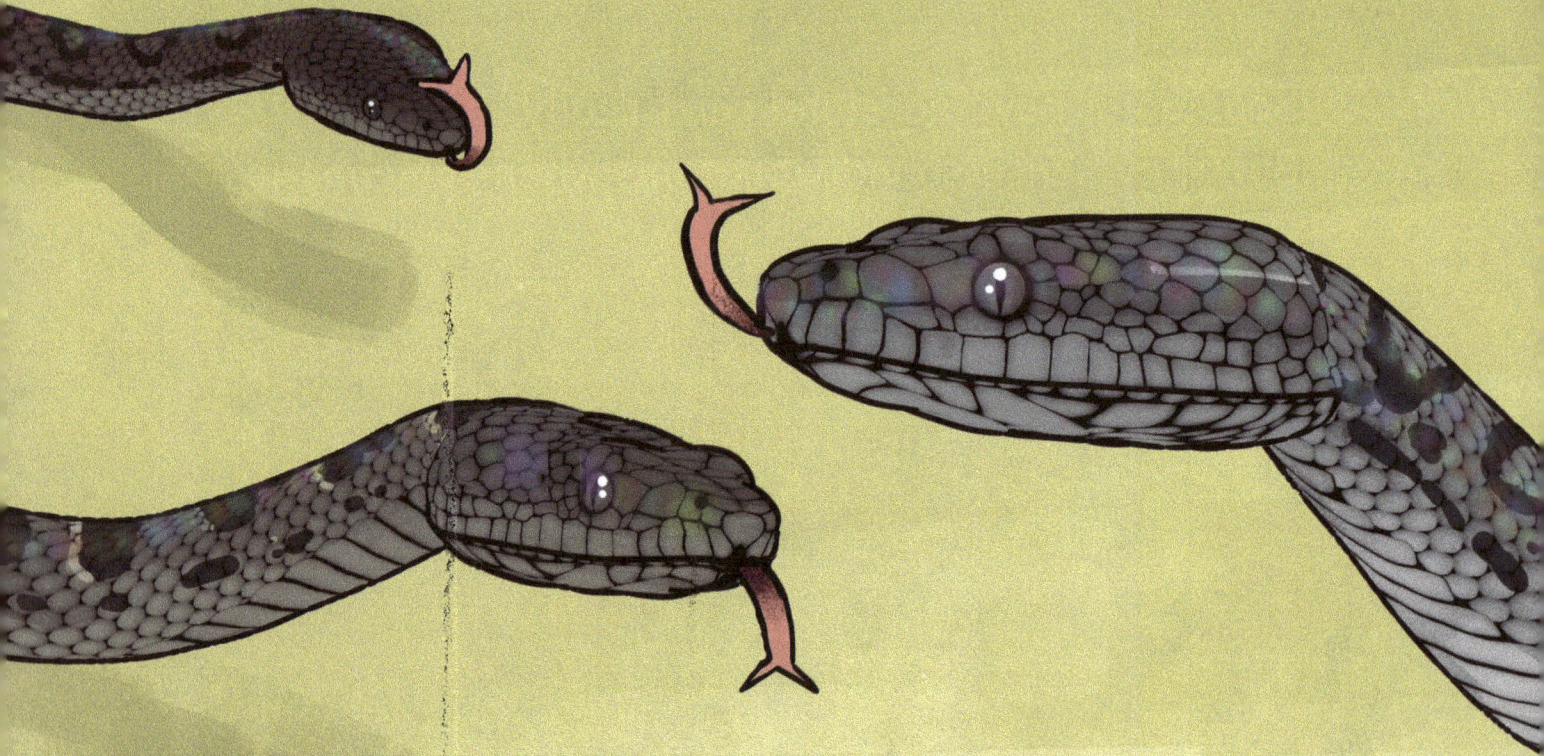

Boas have tongues that they flick out and back.
They're not smelling your scent so they can attack!
They're "tasting" the molecules that float in the air,
Their Jacobson's organs determine what is there.

Bahamian boas can reach eight feet in length;
As constrictors, their muscles have great strength.
The females are larger, and as for the males,
They are smaller, but they have longer tails.

Being a predator, the boa must be bold,
Striking first with its teeth, to grab hold.
Tiny curved teeth have no venom, so boas
    squeeze their prey,
Constricting the victim's chest, suffocating
    them that way.
The jaws separate and the lower jaw moves
    around,
Helping the large meal to slide in and down.

Young boas feed on anole lizards up in trees.
Adults feed on frogs, birds and rats, with
  much ease.
A long-lived boa can eat hundreds of rats -
You should remember this important fact.

Boas are most often seen in summer and spring;
That's the time for hunting and mating flings.

20

In April, boas start looking for love;
Their mate could be on the ground or in branches above.
These snakes breed mostly in the dry season,
So young are born when prey are active - a good reason!

Viviparous boas keep their eggs safe inside,
Not laid in nests - baby boas are born alive!
From one to seventy eggs are incubated;
In the fall, babies emerge, liberated.

For boas, the greatest threat is man,
People kill them on sight because they can.
"I don't care how much rat it kill or harmless it be,
I will kill them if they come around me.
Don't let them live, chop them to death,
And when killing them, hold your breath.
Please kill it before it lays its eggs,
It will eat my children," the person begs.

23

Dogs and cats also kill boas, I'm afraid,
And poachers take them for the illegal pet trade.

24

Although for many years they've been around,
New kinds of Bahamian boas are still being found.
The Silver Boa of Conception is new to the list,
The Crooked-Acklins has now been seen and proved
   to exist.
On these and unknown boas, scientists are working,
Who knows where other wonderful rainbow snakes
   may be lurking.

Now it's time to meet the Bahamian boas here;
Learn about them and see you have nothing to fear.

Meet the Boas

1. Central Bahamas Boa (Chilabothrus strigilatus)

The biggest boa can reach eight feet long;
It's amazing to imagine a snake that strong.
Females are the bigger and heavier,
So a slim boa is a him, not a her.
Found on the Great Bahama Bank islands,
They are the largest predators on the drylands.
Each island's boa is unique, as you will see.
With colors and patterns in great variety.

Central Bahamas Boa subspecies
There are 5 subspecies:
New Providence Boa (Chilabothrus strigilatus strigilatus)

adult

photo © Graham Reynolds

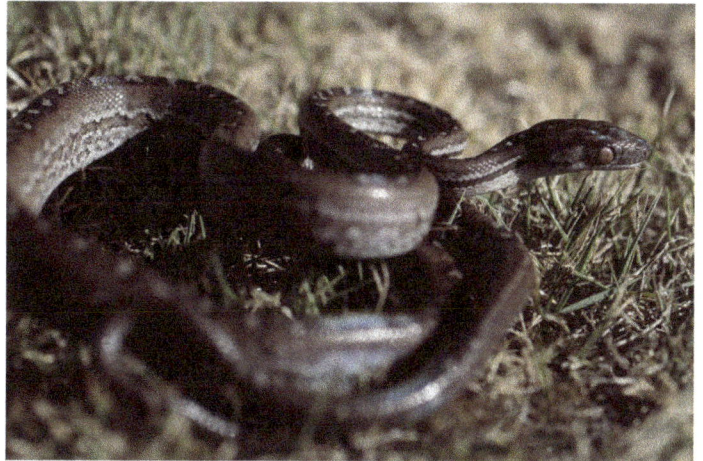

baby (juvenile)

photo © Graham Reynolds

Andros Boa (Chilabothrus strigilatus fowleri)

photo © Graham Reynolds

29

photo © Graham Reynolds

Bimini Boa (Chilabothrus strigilatus fosteri)

Ragged Island Boa

(Chilabothrus strigilatus mccraniei)

photo © Shannan Yates

30

Cat Island Boa (Chilabothrus strigilatus ailurus)

2. Abaco Boa (Chilabothrus exsul)

The smallest boa is only one meter long,
On the island of Abaco is where they belong.
As juveniles, bright orange is their hue,
As adults their grey color is quite subdued.

At night, they hunt for mammals and birds,
But small lizards for food are preferred.
They hide in rocks and logs during the day,
In coppice habitats, the boas prefer to stay.

adult

photo © Graham Reynolds

adult

baby (juvenile)

3. Southern Bahamas Boa (Chilabothrus chrysogaster)

Though on Turks and Caicos they are well known,
They are rarely seen on their Inagua home.
These boas are on the small side,
Often found under rocks, where they hide.
Small birds and their eggs are their usual food,
But they'll eat rodents and lizards when in the mood.

photo © Graham Reynolds

34

4. Silver Boa (Chilabothrus argentum)

On Conception, the newest boa species was found,
Up in the trees, of course, not on the ground.
The boa whose population is most in danger,
Up until a few years ago was a stranger.

The silver color gives us the
snake's name;
That it's in danger is a
terrible shame.

photo © Graham Reynolds

photo © Graham Reynolds

The Silver Boa looks silver by day or night,
The scales on its bottom are a pure creamy white.
They live in trees where birds are their prey;
Prehensile tails coil the branches in case they sway.
Babies have false eyespots on their heads,
Not to see, but to frighten off predators instead.

photo © Graham Reynolds

5. Crooked-Acklins Boa (Chilabothrus schwartzi)

They start out orange in color then turn silver as they age.
A few photos are all we know at this stage.

photo © Alberto Puente-Rolón

Pygmy Boas Tropidophis species
Other snakes have a similar name
But they're not related, which is a shame.
They're also called boas, but don't be confused,
Even if similar names are used.

The pygmy or dwarf boas are very small.
Only a foot or so long – yes, that's all.

The pygmy is called the snake of shame.
How did this snake get such a strange name?
The fearful pygmy curls up, hiding its head,
Then you'll find from its eyes and mouth, it's bled.

Hidden during the day, pygmies are active at night;
For camouflage they change color from dark to light.

photo © Graham Reynolds

1 inch

Like other boas, constriction is their way;
Small reptiles and amphibians are their usual prey.
The pygmies lure with orange tail tips that sway.
After a night's hunt, they sleep through the day.

Young boas have dark spots on a light background,
Adults are a darker color where faint spots are found.
T. curtus is the most common pygmy you'll see.
In the north and center, its sub-species number
    three.
Only on Great Inagua is where *T. canus* will be.
They're ground dwellers, rarely found in a tree.

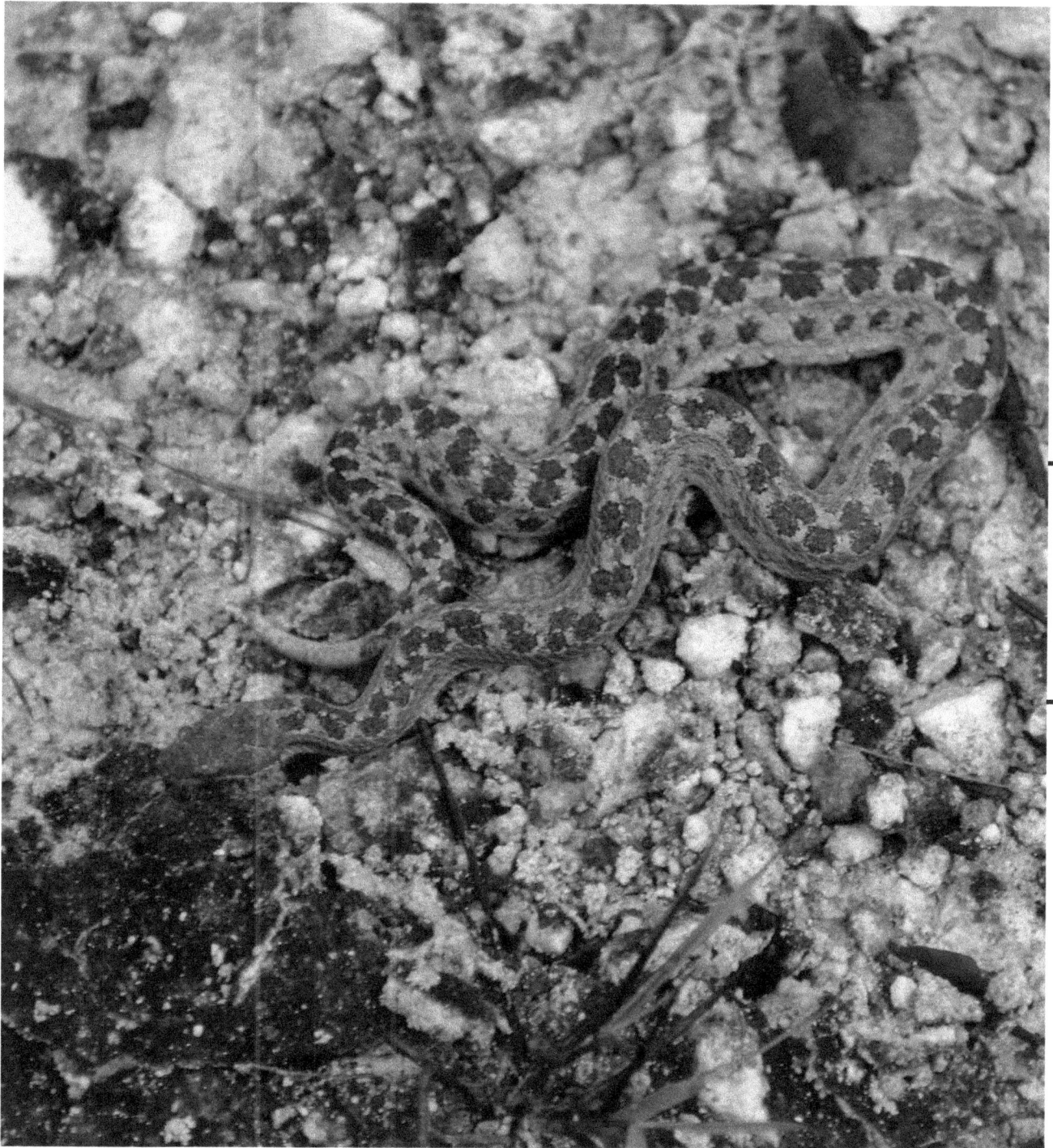

1 inch

Dedication

This book is dedicated to R. Graham Reynolds. I met Graham at a Bahamas National Trust Natural History Conference where he made the exciting announcement of the discovery of another species of Bahamian boa, the Silver boa. Two years later, at another NHC, he announced the rediscovery of a long-lost species of Bahamian boa, the Crooked-Acklins. I wanted to get this book finished before he finds another one! But don't worry, if he does, I'll release an updated edition.

Acknowledgments

My interest in Bahamian boas is due to two scientists, R. Graham Reynolds and Scott Johnson. I owe everything I know about the boas to them, but any mistakes are solely my own.

I couldn't write my books without the support of several fellow authors. Lori Bonati and Susan M. Oyler helped me with the poetry. Brad Peterson, Pamela Bickell and Kate Steele provided valuable insight and editing. The Tucson Poetry Society also provided encouragement. Nora Miller, editor, created the actual book; I couldn't do this without her. I also thank Anderson Atlas for his artwork and his commitment to accuracy.

Biographies

Elaine A. Powers, originally from Peoria, IL, currently resides in Tucson, AZ. After a career as a laboratory biologist, she is now pursuing her dream of writing science-based children's books and murder mysteries as well as continuing her work as a citizen scientist for iguana conservation. Her iguanas and tortoises continually inspire her.

For more information, visit her website www.elaineapowers as well as the publisher's webpage www.lyricpower.net.

More Books by Elaine

Clarissa Catfish Visits the Peoria River Museum

45

Cleo and Tabby: Unexpected Friends

Curtis Curly-tail and the Ship of Sneakers

Curtis Curly-tail Hears a Hutia

Curtis Curly-tail is Lizardnapped!

Don't Call Me Turtle!

Don't Make Me Fly!

Don't Make Me Rattle!

Dragon of Nani Cave

Fly Back to the Brac, Brian Brown Booby

Grow Home, Little Seeds

How (Not) to Photograph a Hummingbird

Lime Lizard Lads and the Ship of Sneakers

Silent Rocks

And several iguana identification booklets and audio theatre scripts

ANDERSON ATLAS

Bradley N. Peterson (pen name Anderson Atlas) is an author and illustrator that lives in the hot Sonoran Desert among scaled survivors, steely eye hawks and majestic saguaros and is inspired by crowded malls, streams hidden by massive boulders, dense forests, and distant mountain ranges.

He went to school for graphic design, but found putting exciting and lengthy stories to his illustrations fit like bananas and ice cream. He's written children's books, YA and middle grade novels and a conspiracy-filled apocalypse novel. He also runs a successful freelance career involving painting for other children's book writers and novelists.

Atlas, his wife, son and daughter live in Tucson, Arizona, where he is actively involved in book groups, helping with a ninja kid class and sailing in Southern Arizona lakes. In his free time, he reads, watches movies and spends a lot of time with family and friends.

His freelance website is BradPetersonArt.com
His Author platform page is AndersonAtlas.com

Tabby, the Five-Finger Fairy, Ambassador for Wildlife Conservation in The Bahamas

Tabby, the Five-Finger Fairy is the creation of Scott Johnson and is a fun and strong character who Scott hopes will become a face for wildlife conservation in The Bahamas. He named her "Tabby" after the scientific name of plant that she originated from, the five finger. A proud Bahamian, Tabby loves Bahamian wildlife, Bahamian bush teas and loves making friends with both animals and humans alike. She is bright, resourceful, funny and tenacious and focuses

49

on important conservation issues that threaten Bahamian wildlife such as wildlife smuggling, habitat loss, invasive species and human intolerance of animals such as snakes and spiders. She lives in a five finger tree and if you look carefully, you may see her sitting comfortably in her tree sipping on a delicious cup of Bahamian bush tea.

LYRIC POWER

www.ingramcontent.com/pod-product-compliance
Lightning Source LLC
Chambersburg PA
CBHW080631030426
42336CB00018B/3150